Marlisa Bernardi de Almeida

Mathematics and problem solving

Marlisa Bernardi de Almeida

Mathematics and problem solving

Problem solving in early childhood education: it's possible, it's fantastic!

ScienciaScripts

Imprint

Any brand names and product names mentioned in this book are subject to trademark, brand or patent protection and are trademarks or registered trademarks of their respective holders. The use of brand names, product names, common names, trade names, product descriptions etc. even without a particular marking in this work is in no way to be construed to mean that such names may be regarded as unrestricted in respect of trademark and brand protection legislation and could thus be used by anyone.

Cover image: www.ingimage.com

This book is a translation from the original published under ISBN 978-613-9-63076-9.

Publisher:
Sciencia Scripts
is a trademark of
Dodo Books Indian Ocean Ltd. and OmniScriptum S.R.L publishing group

120 High Road, East Finchley, London, N2 9ED, United Kingdom
Str. Armeneasca 28/1, office 1, Chisinau MD-2012, Republic of Moldova, Europe
Printed at: see last page
ISBN: 978-620-7-76630-7

SUMMARY

1 INTRODUCTION

Concerns about quality mathematics teaching from nursery school onwards are becoming more and more frequent, and there are countless studies that suggest ways of ensuring that students in this age group have the opportunity to start their first contacts with this subject in the right way.

For this reason, a Mathematics work proposal for Early Childhood Education should encourage the exploration of a wide variety of not only numerical ideas, but also those relating to geometry, measurements and notions of statistics, so that children develop and maintain a curiosity about Mathematics with pleasure, acquiring different ways of perceiving reality.

Working with Mathematics in Early Childhood Education must incorporate the contexts of the real world, the experiences and natural language of the child in the development of mathematical notions, without forgetting that the school must make the student go beyond what they seem to know, trying to understand how they think, what knowledge they bring from their experience in the world and making interferences in order to lead each student to progressively expand their mathematical notions. There are several possible paths to follow when you want to organize a proposal for Early Childhood Education with these concerns.

However, the most recent studies that support pedagogical practice have pointed to problem solving as an efficient methodological perspective for teaching mathematics, especially in early childhood education.

Authors such as IMENES, POLYA, NASSER and SMOLE, who advocate problem solving, show that this is the basic activity of doing and thinking in mathematics, which justifies the need to learn specific concepts and procedures in this area of knowledge.

However, observing the pedagogical practice in Mathematics classes in Early Childhood Education shows that when problem solving is used, it is most often practiced in a dry, unmotivated, mechanized way, where children are limited to mere training and are not "prepared" to think of solutions. Another issue that arises is that the teacher thinks he or she is a "problem solver", which creates a certain dependency and insecurity in the students when it comes to finding strategies to solve problems.

In this way, problem-solving completely deviates from its "main destination", ending up on other paths that only lead to despair and failure for the student, the teacher and consequently the whole school.

Problems have not played their true role in teaching, especially in Early Childhood Education (five to six years old) and the early grades of Primary School (1st year of the 1st cycle). For the vast majority of students, solving a problem means doing calculations with the numbers in the statement or applying something they have learned in class without at least interpreting and analyzing it; the problem thus becomes a disguise for the mere application of operations, formulas and techniques.

Mathematical knowledge is not presented to the student as a system of concepts that allows them to solve a set of problems, but as an endless symbolic discourse, abstract and incomprehensible, causing students to have a certain aversion to Problem Solving since Kindergarten.

By analyzing the real situation of Problem Solving in the classroom, we sought to carry out this research in order to propose significant changes in teaching practice. As IMENES (2002, p. 34) says: "The biggest mistake in math classes is to spend 95% of the time doing math problems. Teaching should focus on problem-solving."

We want children to think, to have their own ideas, to exercise their reasoning skills. In order to achieve this goal, we must make the most of every opportunity and especially work with problem-solving, which is the best way to do this.

The research will address Problem Solving by investigating how this methodological alternative is being used in the classroom, emphasizing the strategies that students use when faced with really challenging problems. Its main objective is to analyze how Problem Solving as a methodological perspective at the service of teaching and learning Mathematics broadens the purely methodological vision and overturns the issue of the great difficulty that students and teachers face when Problem Solving is proposed in Mathematics classes in Early Childhood Education.

In addition, this research meets the pedagogical and mathematical needs of students as well as teachers, as it is an instrument of support and a basis for the work of those who believe that learning is a dynamic, reflective act, led by the child's curious and questioning gaze.

This research will show what kind of problem-solving experience children should have and that even before they are readers, children are already capable of solving problems in math classes. And it goes further, showing the importance of working with Problem Solving right from Early Childhood Education.

The first part of the work takes a theoretical approach that underpins the research, bringing reflections and considerations on the importance of Problem Solving since the dawn of humanity, looking back and showing that well-planned work with Problem Solving is essential from Early Childhood Education, in order to have citizens prepared to act in the society of the third millennium.

It also discusses some issues pertinent to the teaching-learning process in a constructivist conception, investigating factors that influence this process in Problem Solving. The theoretical reflection that this work proposes highlights the teacher's attitude in the classroom, emphasizing a problem-solving approach so that positive results can be achieved with the proposal presented.

The second part of the research presents a proposal for Early Childhood Education teachers - of children aged between five and six - to take on Problem Solving as a methodological alternative for teaching Mathematics that should permeate the teaching-learning process, making it necessary to reshape teaching practice and the teacher's own attitude towards problems. To help fulfill this proposal, there are practical suggestions for working with problems, in order to equip teachers to really take it on and make the teaching of mathematics really challenging.

Finally, the last part of the work proves, through practice, that it is possible to work with problem solving in Early Childhood Education and that they evolve much faster than one might imagine when faced with really challenging problems. Field research is used to show how students perform when faced with problems, to identify the strategies they use, to analyze the role of orality, pictorial representations and writing as teaching resources, allowing teachers to see a new dimension to school practice.

It is hoped that this research can make a significant contribution to improving the quality of Mathematics teaching in Early Childhood Education with children aged five to six, leading all those who have access to it to realize that solving problems is one of the main reasons for studying Mathematics.

It is believed that, after reading this work, teachers will be better prepared and enthusiastic about the stimulating task of teaching creatively, as the aim is to awaken their sensitivity to a new charm that can emerge at school and which springs essentially from encouraging children to embark on the adventure of formulating and solving problems. Fantasy, irrelevance and mistakes will be part of this adventure, but the teacher, before being moved to be patient with this, would perhaps do better to let himself be involved by the "magical" force that problems can awaken in everyone.

2   PROBLEM SOLVING IN THE HISTORY OF MATHEMATICS

Looking back over the history of mathematics since the Babylonians and Ancient Egypt, great importance, because it appears clearly, is attributed to the problem. Much of ancient mathematics has been handed down to us in the form of problems.

The historical context also points to the importance of using the problem not as an end in itself, but one that is linked to everyday reality.

Mathematical thinking is characterized by problem-solving activity, i.e. the tendency of human beings (especially children) to ask questions and seek answers. Therefore, basic mathematical notions must be based on and constructed from concrete problem situations, which come from the children's experiences and which provide an opportunity to look at previous mathematical knowledge, the strategies used, as well as the difficulties faced by the child.

It is known that the mathematical tradition prior to the Greek period was based on problem-solving, since we don't find theorems being solved, but rather problems accompanied by how to solve them. These problems are generally related to the reality and the subjects they were designed to deal with: food (such as the bread and beer of the Egyptians), finance, or even the division of land between heirs. These problems were presented to make them attractive and educational.

In the Rhind papyrus, in particular, which dates back to around 1600 BC, it can be seen that most of the eighty-seven problems presented are of a practical nature, although some of them are intended as riddles or mathematical games.

Leonardo Pisano (FIBONACCI), in his work *Liber Abbaci,* written in 1202, uses the same style of problems, presenting more than one solution process. Another aspect to note in FIBONACCI's work is the possibility that in some cases the problem is not solvable due to the conditions proposed or for certain values (meaningless solution). It is not uncommon to find algebraic or geometric problems, using diagrams, graphical representations and tables.

Most works from the Middle Ages and the Renaissance are set up with situations involving problems. There are both practical and recreational problems, the wording of which is still used today. One example is problems involving weighing a limited number of weights.

In an earlier period we find *the Propositiones* by D'Alcuin, who is thought to have lived at the court of Charlemagne. The *Propositiones* are questions to make young people "more intelligent", which influenced textbook authors for several centuries.

According to BARON (1985), DESCARTES, LEIBNITZ, NEWTON and EULER paid special attention to Problem Solving, since all of their works presented their theories based on a problem that they had to find a solution to.

We can see that Mathematics has advanced significantly through the scientific investigation proposed by problems.

2.1 PROBLEM SOLVING IN RECENT DECADES

Although the history of mathematics before Christ shows the importance of solving problems, the subject of problem solving has only been much discussed and analyzed in recent decades and especially at the beginning of this new millennium, among teachers and educators as well as researchers and curriculum developers.

The proposal to put Problem Solving at the forefront of school Mathematics has been put forward insistently for some years now by various researchers, because for them it is the driving force behind the development of Mathematics and its activity. GEORGE POLYA's famous book, *How To Save It, is* considered a milestone, as it was first published in 1945 and has since been reprinted several times and has served as an inspiration for many authors.

POLYA (1977) focuses on Problem Solving as the process of applying previously acquired knowledge to new situations. POLYA's concept gained greater importance in the 1970s, when educators began to focus their attention on the processes or procedures used by students to solve problems. The implications for teaching became a focus on the procedures or steps used to arrive at the answer, while the answer lost its importance.

At the end of the 1970s and during the 1980s, Problem Solving became a basic skill, with clear indications that all students should learn to solve problems and that careful choices need to be made about the techniques and problems to be used in teaching. From this perspective, it is necessary to consider the problems that involve specific content, the different types of problems and the methods of solving them in order to achieve mathematical learning.

In 1980, the *National Council of Teacher of Mathematics*, the recognized American association of mathematics teachers, dedicated its publication to Problem Solving, reinforcing curricular proposals established in the United States that indicated that Problem Solving would be the focus of teaching and research in the 80s.

More recently, in the 1990s, Problem Solving took on a new dimension, being described as a methodology for teaching mathematics and, as such, it became a set of strategies for teaching and developing mathematical learning. This conception is present in the challenge that can trigger teaching through projects using problematization. It is also present in broader guidelines for teaching Mathematics, which correspond to lines of research and action in Mathematics Education.

Based on the influence of all these concepts and the action research carried out with teachers and students, the second chapter defines what is meant by Problem Solving.

2.2   THE CURRENT ROLE OF PROBLEM SOLVING

Today, in order to perform effectively in this new century, students will need a wider range of mathematical knowledge, and must develop their problem-solving skills and higher-order thinking.

That's why it's clear that problem-solving is still the main axis of mathematics teaching, right from nursery school, because essentially, doing mathematics is about finding solutions to problems, and it's through these solutions that we arrive at new discoveries that are so necessary in the history of humanity, especially today, when transformation is very rapid and requires discoveries all the time.

These rapid social changes and the ever-increasing and ever-faster improvement of technology make it impossible to predict exactly which mathematical skills, concepts and algorithms would be useful today to prepare a student for their future life. Teaching only concepts and algorithms doesn't seem to be an efficient way forward, as they may become obsolete in a few years' time, when the child is at the peak of their productive life. Therefore, a very reasonable course of action is to prepare children to deal with new situations, whatever they may be. To do this, it is essential to develop initiative, an exploratory spirit, creativity and autonomy through Problem Solving.

Today's society needs people who are active and involved, who have to make decisions quickly and as accurately as possible. It is therefore necessary to train mathematically literate citizens who know how to solve their problems in commerce, economics, administration, engineering, medicine, weather forecasting and other areas of daily life intelligently and effectively. For this to happen, children need to have Problem Solving as a substantial part of their Mathematics curriculum, so that they can develop their ability to deal with problem situations from an early age.

Students are expected to be able to solve problem situations, validating strategies and results, developing forms of reasoning and processes such as deduction, intuition, analogy, estimation and using mathematical concepts and procedures, as well as available technological tools, so that they are able to face the job market and the globalized world.

In addition, students need to interact with their peers in a cooperative way, working collectively to find solutions to proposed problems, identifying aspects that they agree or disagree with when discussing a subject, respecting their colleagues' way of thinking and learning from them, as this is one of the keys to entering the highly competitive job market: knowing how to cooperate, but having the necessary knowledge and skills to be "the best" together with their team.

It is clear to see the great importance that should be given to Problem Solving, because if we want to train citizens who are active and active in society, it is urgent that they know how to solve problems, and this is only possible through Problem Solving, not just in the final grades of primary or secondary school, because the world doesn't just need concepts, not least because the Internet and books make these available at any time, but solutions, strategies, new ways of thinking and solving the situations that arise in everyday life in this constantly changing world.

3   THE CONCEPT OF PROBLEM SOLVING

In the absence of a suitable term - and in order not to use others that might give a reduced idea of this conception of Problem Solving - it will be called: *methodological perspective, a* term suggested by SMOLE & DINIZ (2001).

This means that Problem Solving corresponds to a way of organizing teaching which involves more than purely methodological aspects, including an attitude towards what it means to teach and, consequently, what it means to learn. That's why I chose the term "perspective", which means "a certain way of seeing" or "a certain point of view", to broaden the conceptualization of Problem Solving as a simple methodology or set of didactic guidelines.

Firstly, Problem Solving is based on proposing and dealing with problem situations. In other words, broadening the concept of problem, we should consider that Problem Solving deals with situations that have no obvious solution and that require the solver to combine their knowledge and decide how to use it in search of a solution.

This perspective breaks with the limited vision of problems that can be called conventional and which are traditionally proposed to students. When conventional problems are adopted as the only material for working with Problem Solving at school, this can lead students to become fragile and insecure when faced with situations that require a greater challenge. When faced with a problem in which they can't identify a model to follow, all they can do is give up or wait for an answer from a classmate or teacher. Often, they will solve the problem mechanically, without understanding what they have done and without trusting the answer they get, being unable to check whether the answer is appropriate to the data presented or to the question posed in the statement.

Thus, the first characteristic of the methodological perspective of Problem Solving is, according to SMOLE (2001, p.17), "...to consider as a problem any situation that allows for some problematization", that is, that allows for some questioning or investigation. These situations can be planned activities, games, searching for and selecting information, solving unconventional or even conventional problems, as long as they allow for an investigative process and trigger in the child the need to find a solution using the resources available to them at the time.

Problem-solving is, according to DANTE (1996, p.15):

> "...a way of developing class work, it is a methodological perspective through which students are involved in doing mathematics, that is, they become capable of

formulating and solving mathematical questions on their own and through the possibility of questioning and raising hypotheses, they acquire, relate and apply mathematical concepts."

The problem-solving approach is characterized by a posture of nonconformity in the face of obstacles and what has been established by others. It is a continuous exercise in the development of critical thinking and creativity, which are key characteristics of those who make science and mathematics objectives.

From this point of view, solving problems in Early Childhood Education is a space for communicating ideas, making statements, investigating relationships and gaining confidence in their learning abilities. It is a time to develop notions, procedures and attitudes towards mathematical knowledge. A problem-solving approach helps students to make more sense of the concepts, skills and relationships that are essential to the mathematics curriculum at this level.

That's why Problem Solving shouldn't be seen as a separate subject, an isolated topic in the curriculum, but rather a methodology that should be used throughout the teaching and learning process. It represents much more than teaching students to use operative techniques or algorithmic procedures; it involves getting them to activate their network of knowledge, make connections, establish links between topics in Mathematics and between other areas of knowledge.

Problem-solving represents a process of investigation in which all the student's knowledge is combined, associated, so that they can creatively and autonomously solve a problem situation in any area of knowledge. In this approach, students must be questioned at all times and asked to defend, argue and justify their ideas, even if the solutions are not the most convincing. According to SMOLE (2000, p. 21) "Students need to be encouraged to evaluate their own response, their own problem, turning it into a source of new problems".

The fact that the student is encouraged to question their own answer, to question the problem, to transform a given problem into a source of new problems, shows a concept of teaching and learning that is not merely about reproducing knowledge, but about reflective action that builds knowledge. In this way, it seeks to demystify all the drama and aversion to problems, making students develop their reasoning and creativity, motivating them to learn and enjoy mathematics. In addition, no one will feel alone, wrapped up in thoughts and doubts...

A problem only exists when the student is led to interpret the question and to structure and contextualize the situation presented; the solution is not available at the outset, but is constructed. Being faced with a problem allows students to work out one or more solution procedures, compare the result with that of their classmates and validate their procedures.

Through Problem Solving, students learn to work in groups, to confront ideas and solutions, to defend their hypotheses and also to accept and question opposing hypotheses. Federal deputy and educator Ester Grossi says of a good educator: "Let them be aware that everyone can learn mathematics and that this happens through solving problems". (GROSSI, 2001, p. 37)

According to POLYA (1977, p. 49):

> "Solving problems is the specific achievement of intelligence and intelligence is the specific gift of man. Solving problems characterizes human nature. Therefore, learning to solve problems is the main reason for studying mathematics, and this point of view influences the nature of the entire mathematics curriculum and has important implications for classroom practice."

Considering Problem Solving as a goal can influence everything that is done in the teaching of Mathematics, showing another proposal for teaching. This requires planned and constant work, using many and varied sources of problem solving, from those that arise in the students' daily lives to more elaborate proposals that the teacher can make.

Problem-solving, aimed at building knowledge and developing reasoning and self-confidence, is one of the most effective methodologies for teaching mathematics, as it covers a wide range of skills, potential and knowledge in a wide variety of areas, especially mathematics.

3.1 SOLVING MATH PROBLEMS IN EARLY CHILDHOOD EDUCATION

Is it important to develop Problem Solving in Early Childhood Education?

What does it mean for a child in kindergarten to solve a problem?

Could it be that, even before becoming a reader, a child is already capable of solving problems in math class?

What kind of Problem Solving experience should children have?

These are some of the questions with which it is necessary to reflect on the practice of teaching Mathematics in Early Childhood Education.

It is well known that one of the main reasons for studying mathematics at school is to develop problem-solving skills. This skill is important not only for children's mathematical learning, but also for the development of their potential in terms of intelligence and cognition. For this reason, we believe that problem-solving should be present in Mathematics teaching from Early Childhood Education, not only because of its importance as a way of developing various skills, but especially because it gives students the joy of overcoming obstacles created by their own curiosity, thus experiencing what it means to do Mathematics.

For a child, just as for an adult, a problem is any situation they face and can't find an immediate solution that allows them to link the starting data to the objective they want to achieve. The notion of a problem implies the idea of something new, something that has never been done before, something that is not yet understood.

This perspective of Problem Solving has the characteristic of broadening the concept of problem and, as a consequence, knowing how to problematize. The questions asked depend on the objectives to be achieved. This may seem obvious, but it is common to find the idea that problematizing means subjecting children to a list of questions formulated by the teacher, but without being clear about what the student is trying to develop. Therefore, when practicing Problem Solving, it is essential to carefully plan the activities and the way in which the questions are asked.

It is also necessary to explain what is meant by the content to be worked on in Early Childhood Education in order to understand many of the suggestions with Problem Solving that will be developed in this research.

Content is understood to include, in addition to the specific concepts and facts of mathematics, the skills needed to ensure that children are trained to be confident in their knowledge and capable of beginning to understand certain procedures in order to use them properly. Within the idea of content are also the attitudes that enable learning and that contribute to forming citizens.

In Early Childhood Education, according to current teaching proposals and given the very different characteristics of this age group, the objectives include the development of skills and attitudes as well as concepts that complement this training, but are not at the heart of it.

This shows that Problem Solving has another important characteristic, which is the non-separation between content and form. In other words, there is no teaching method without working on some content, and all content is closely linked to one or more forms of approach.

Thus, problematizations must aim to achieve some content that deserves to be taught and learned. This is why it is so necessary to plan and choose content and activities, because they must contain good questions that encourage students to learn while they search for answers.

Another aspect that deserves attention is that problematization includes what is called a metacognitive process, i.e. when you think about what you have thought or done.

Each new question requires a return to what you know in order to face the challenge. This requires a combination of knowledge and a more elaborate form of reasoning.

In order to develop all these skills, children need to be challenged to solve problems in specially designed situations from the start of school.

In addition, teachers should be clear that the mathematical ideas and procedures that children develop in childhood support the mathematics they will study later on and that the early school years can develop attitudes towards mathematics, making them believe in their ability to learn. So it is with Problem Solving, because the development of a positive attitude towards facing and solving problem situations will influence children's future success in this activity.

The absence or secondary place given to problems in the teaching of mathematics to children who are not yet literate has long-term consequences. Often, these activities are not approached systematically, or are only planned from the second or third grade onwards. For many primary school teachers, problem-solving is therefore a difficult task that is poorly understood by the students. The questions students often ask the teacher include: *What do I have to do? Is it right? Which account to solve? Is it more or less?* reflect these difficulties.

Some recent studies carried out by GEPEM (Group for Studies and Research in Mathematics Education) on the behaviour of students faced with unusual problems show that they have built up a representation of what a math problem is characterized by some mistaken principles, such as: *a problem has numbers and all numbers must be used, a problem always has a solution, there is only one way to solve a problem.* Behavior linked to these representations of problems can be observed as early as elementary school, when students have several operations at their disposal, but reduce the solution to just one to be carried out with the first two numbers found.

When working with Early Childhood Education, it is clear that many teachers have this idea of what a problem is, because they were trained in a teaching model centered on the problems proposed in textbooks. When the concept of a problem is broadened to include a problem-situation, the aims of Mathematics in Early Childhood Education are no longer just to count and write numbers, and some of the existing beliefs among teachers, which have led to the lack of attention given to problems, fall by the wayside.

The first of these beliefs is that *children need to be readers in order to solve problems.* This argument is easily refuted, as not knowing how to read or write is not synonymous with an inability to listen, speak, understand or think. In addition, children, even non-readers, are challenged by many situations on a daily basis, which they face and solve with ease.

From the perspective of Problem Solving, the teacher is a reader who can put his or her skills to work for his or her students, and working with problems is one of the elements that triggers the acquisition of reading and writing in students in the literacy phase. By working with problem texts, you can explore words, letters, rhymes, among many other activities typical of the literacy process.

A second widespread belief is that *in order to solve problems properly, children need to master numerical concepts.* This belief is even more unfounded than the previous one, as it is often possible to problematize situations in the classroom and the problems they encounter in everyday life, as well as in Mathematics itself, are not necessarily numerical.

A third reason for the secondary position of Problem Solving in Early Childhood Education lies in the argument that *in order to solve problems, children first need to have some knowledge of mathematical operations and signs.*

Most teachers think of problems as applications of operative techniques, rather than a starting point that can lead to a calculation.

KAMII (1995) states that the teaching of calculation techniques which precedes the presentation of verbal problems in most textbooks, rather than meaningful situations for the child, is a manifestation of the conviction that without these techniques children will not be able to reason arithmetically. "Arithmetic is not born from technique, but from the child's ability to think logically." (KAMII, 1995, p. 45)

So instead of thinking of problems as being about this or that operation, we should think of them as questions that children try to answer by thinking for themselves. In this way, it requires nothing more than every child's natural ability to be fascinated by challenges.

This can be seen in an example of a six-year-old student solving a problem that can be considered conventional and numerical.

Using several little fish, she chose four, according to the problem, and drew three little worms inside each one. She counted all the minnows and found that there were twelve. Although she didn't know how to represent the numeral 12 arithmetically, she managed to arrive at the correct answer through drawing and oral counting, but with the help of the posters in the classroom she wrote the numeral correctly.

We can also see that she doesn't know how to read or write conventionally either - as we can see from her writing of the word *minhoquinhas (MOIHPRFS),* which according to the studies carried out by FERREIRO (1985) is in the pre-syllabic II stage, using the initial and final letters of the word and relating them to its sound value - even so, she managed to solve the proposed problem satisfactorily.

This and many other examples that will emerge throughout this research will show the importance and magic of working with Problem Solving in Early Childhood Education and that this methodological perspective breaks with all the beliefs explained above.

This is why reviewing these conceptions is one of the first steps educators can take to improve the approach to problems in their classes. It is also important to point out that Problem Solving should not be restricted to simply instructing how to solve a problem or certain types of problems. It is also not a question of considering Problem Solving as an isolated content within the curriculum. Rather, it is a methodological perspective, an orientation for mathematical learning, because it provides the context in which mathematical concepts, procedures and attitudes can be learned.

4 A PROBLEM-SOLVING PROPOSAL FOR EARLY CHILDHOOD EDUCATION

This third and fourth chapter will deal with a proposal for working with Problem Solving in the classroom. This proposal provides a theoretical basis as well as practical suggestions on how to develop this work in the classroom.

4.1 PROBLEMATIZING APPROACH

With this approach to classroom work, the aim is to exercise reasoning at every opportunity. It should take place in all lessons and not just in "Maths lessons". The reason is that you can't just expect students to reason on the day they work on math problems.

A problem-solving approach consists of treating the most diverse subjects as if they were problems. In other words, math problems are not the only ones that deserve attention.

Everyday problems at school, in the classroom, for example: *Children run around a lot at playtime and end up hurting themselves too much. What can be done to solve this problem?* Starting from a real situation, the children will be tracing paths, making decisions, planning solutions and checking the validity and practicality of their responses. This is problematizing, this is getting students to think!

So, how can you help them solve their problems, suggest ways forward and reason? The answer lies in the exchange of ideas, in dialogue.

Dialogue enhances the student's reasoning, because they are constantly listened to by the teacher and their classmates. As a result, students feel stimulated to think.

Normally, an ideal way to complement a class of problems solved in groups is to ask a few students to explain to the class how they solved one or other of the problems. This way, once again, the students' reasoning is valued and everyone learns from each other.

Dialogue is not only present when problems are solved in groups. It is the basis of any problem-solving approach. Whenever the teacher asks questions, listens to the students and allows them to present their ideas, dialogue is promoted and reasoning is stimulated.

That's why the teacher should always ask questions, give students opportunities to think and interact with their classmates. They shouldn't be anxious and explain the answer, but rather let the students explain, allow them to speak, and then make the necessary collective interventions.

In this problem-solving approach, the teacher's intervention is necessary in order to question, make the student doubt their answers or prove that they are correct, provoke challenges and "entice" them to solve problems. 4.2 THE TEACHER'S ROLE IN PROBLEM SOLVING IN THE CLASSROOM

According to several teachers, they are so anxious to teach that they don't wait for the child to reason and try to solve a problem, most of the time they just give them the way and the steps to solve it, in other words, the ready-made answer. In this way, they don't give the child the opportunity to associate the new content with what they already know, so that they can build their learning from what they know.

The early grades are where the greatest "learning difficulties" are to be found - despite teachers' concern for their students' learning - their lack of understanding of their students' development leads them to make mistakes, proposing problem situations at inappropriate times or in an inappropriate way.

For learning to take place, children need to understand how to solve and not just systematize the development of problems. However, as a rule, only one answer is taken into account and the discovery of other solutions is not valued, so students are not stimulated to think and reason. Students should be prepared to face everyday problems, and their reasoning and creativity should be encouraged.

DEMO (2001, p. 53) says: "The good teacher is not the one who solves problems, but the one who teaches students how to problematize." This is the key idea for the role of the teacher within this Problem Solving proposal.

PIAGET (apud AZENHA, 1993, p.65) said something that still scandalizes many people today: "If you answer a child's question, you prevent them from learning." This is another precious idea. However, in Brazil, it is believed that the teacher exists to answer questions. According to PIAGET (apud AZENHA,1993), we have to live with doubt, because it is at the heart of learning. This doesn't mean accepting every mistake, but knowing that the teacher can't think for the student. When the teacher instead encourages the search for arguments and counter-arguments, they develop a number of things, from autonomy to respect for others, because you have to stop and listen to your colleague's argument. Learning to criticize others on a logical basis or to accept criticism, all of which is a huge exercise in citizenship.

From a constructivist point of view (AZENHA, 1993), teachers should let their students "work it out" rather than offering a solution. This is what the Japanese do, who practice argumentation brilliantly. A problem is presented and the students try to find the solution through different routes. Sometimes, those who haven't managed to find the solution discover an interesting route, which can be exploited didactically.

According to DEMO (2001), it is difficult for a Brazilian teacher to act like this, not least because he or she has two major problems, for which he or she is not to blame. Firstly, because the Pedagogy faculties are very weak, as they don't work in depth on the issue of learning or focus on the three major axes, formed by the teaching of Philosophy, Language and Mathematics. Secondly, teachers are extremely poorly paid and demotivated. "...with what he earns, he can neither inform himself nor learn." (DEMO, 2001, p. 13)

That's why teachers need to not be discouraged by the obstacles to their training, but always seek knowledge and theories that will help their teaching practice and help them form a posture aimed at being, as DEMO (2001, p. 13) states "...a problem solver and not a problem solver."

First and foremost, taking on Problem Solving in the classroom is not just a change in pedagogical practice, but essentially a change in the way teachers view Problem Solving in their own lives and consequently in their school practice.

4.3 ORGANIZING THE CLASSROOM ENVIRONMENT WHEN WORKING WITH PROBLEM SOLVING

Class work is very important in developing the methodological perspective of Problem Solving, because it is here that meetings, exchanges of experiences, discussions and interactions take place between the children and the teacher. It is also here that the teacher observes the students, their achievements and their difficulties.

In this way, children need to feel that they are participating in an environment that makes sense to them, so that they can engage in their own learning. The classroom environment can be seen as a workshop for teachers and students, and can become a stimulating, welcoming space for serious, organized and joyful work.

It is known that while living in an environment in which they can act, discuss, perform and evaluate with their group, children acquire conditions and experience situations that are favorable to learning. For this reason, the classroom should be a cooperative and stimulating environment for students' development, as well as providing interaction between the different meanings that students will learn or create from the proposals they make and the challenges they overcome. In this sense, working groups become indispensable, as do different teaching resources.

It is a positive environment that encourages students to propose solutions, explore possibilities, raise hypotheses, justify their reasoning and validate their own conclusions. Thus, in this environment, mistakes are part of the learning process and should be explored and used to generate new knowledge, new questions, new investigations, in a permanent process of refining the ideas discussed.

The role of communication between those involved in the class work process is emphasized. Communication defines the situation that will give meaning to the messages exchanged; however, it is not just about transmitting ideas and facts, but mainly about offering new ways of seeing these ideas, of thinking and relating the information received in order to construct meanings. Exploring, investigating, representing their thoughts and actions are communication procedures that should be implicit in the organization of the working environment with the class.

Precisely because representing, listening, speaking, reading and writing are basic communication skills that are essential for learning any content at any time. It is suggested that the work environment should include times for reading and producing mathematical texts, group work, games, making pictorial representations, reading and making books by the children. By varying the processes and forms of communication, the possibility of meaning for an idea that arises in the context of the class is widened. When a child's idea is highlighted, it provokes a reaction from the others, forming a network of interactions and allowing different skills to be mobilized during the discussion.

The work of the teacher according to GARDNER (1994, p. 25):

> "...it's not about solving problems and making decisions alone. He animates and maintains the network of conversations, as well as coordinating the actions. Above all, he tries to discern, during the activities, the new possibilities that could open up to the class, guiding and selecting those that will bring the students closer to the objectives set and to the search for new knowledge."

4.4   PLANNING AND EVALUATING WORK WITH problem SOLVING

The first concern observed in education teachers

It is recommended to start with the simplest proposal, which is the oral problematization of situations close to the student.

Situations involving the distribution of materials in the classroom, making decisions about how to organize themselves for an activity, simple rule games, problematizing based on an image or picture are some of the ways of proposing the first problems, with orality being the first resource for communicating the problem and for the students to present their hypotheses and solutions.

Once children have become familiar with problem situations in simpler language, as they gain confidence in their ways of thinking and look for more precise strategies to communicate their thoughts, their ways of proposing and solving problems can become more elaborate.

In order to develop all the skills involved in the process of solving problems in a way that complements the development of language, socialization, knowledge of oneself and the space around the child, it will be necessary to plan actions to organize both the types of problems proposed and the classroom dynamics in an alternating manner.

In this sense, in order for Problem Solving to be characterized as an action of engagement in the search for a solution to a situation, with the confidence and freedom to choose their way of thinking and reporting this resolution, problem situations can be chosen from among those involving numbers, counting and notions of operations, as well as non-numerical situations. This should be done so that the problem-solving is not restricted to the more conventional situations or those that focus the work only on the development of numerical or arithmetical concepts.

According to DANTE (1996, p. 06) "Collective work alternated with organization in small groups, pairs and even individually, generates different forms of relationship between the students and between them and the teacher." And this collective work allows students to develop more fully during the solving process. That's why the choice of class organization will also depend on the objectives of the activity and the characteristics of the group of students.

Another observation, of a general nature, concerns the proposition of problems to non-reading students through the teacher's reading. If the student doesn't read, the teacher can read the problem to them and suggest that they find a way to express the solution on a piece of paper. It is also possible for a student in the class to read the problem and for everyone to discuss it orally.

There are many possibilities and the teacher will certainly find other ways of working with problems with non-reading children. However, a word of warning is in order. When reading a problem to the class, the teacher must be careful not to emphasize certain words. Because according to IMENES (2002, p.09) "...emphasizing key words can lead to difficulties and errors on the part of the students."

Therefore, when reading a problem to or with children, care must be taken to ensure that the reading is unbiased, i.e. the teacher must not try to facilitate the process, but must provide elements with which they can search, investigate, analyze and, on their own, find the solution to what has been proposed.

Another observation about reading problems is that often, when reading or listening to a problem, students find it difficult because they don't know the terms or words that appear in it. Doubts about this can be overcome by using some strategies on the part of the teacher, such as: raising unknown words with the students, making a list and putting the corresponding meaning next to each one; dramatizing the problem; getting the class to read the problem more slowly.

With this well-planned work, it is believed that the children will form habits of thought that will allow them to gain autonomy in Problem Solving while at the same time advancing in their understanding and mastery of reading processes.

For planning to be truly effective, it must have an evaluation system. This means that it allows for constant and frequent reviews, so that it is possible to make adjustments and adaptations to the plan itself or even correct it during the course of working with Problem Solving.

According to MASETTO (1998, p. 80):

> "...if it is not supported by a diagnostic, formative and reflective evaluation system, planning will lose its flexibility, dynamism, reliability and effectiveness, and could seriously compromise the educational action it is intended to make possible."

Planning is therefore an indispensable tool for teachers, as it allows them to adapt to the time available, select activities, techniques and strategies and evaluate according to the objectives set and within the existing limits.

4.4.1 Planning: the First Step to Practicing Problem Solving

First of all, it's important to point out that all planning only exists because of one or more objectives that you want to achieve. That's why you need to know what the objectives of Problem Solving for Early Childhood Education are.

According to the National Curriculum Guidelines for Early Childhood Education (RCN'S, 1997) the skills, procedures and attitudes that children are expected to develop in the process of Problem Solving are as follows:

S Reading and interpreting different types of texts.

S Develop and use oral, pictorial and written languages.

S Arguing and questioning.

S Raise hypotheses.

S Check the hypotheses made.

S Check that the answers given are appropriate to the situation or question.

S Make different representations of the same situation.

J Apply the mathematical knowledge involved in problem situations.

S Relying on your own knowledge.

S Listening to and respecting others.

S Persevere in finding a solution to a situation.

S Working cooperatively.

In order to achieve all of these objectives, planning should ensure that problem-solving is used throughout the school year, with alternating proposals. Thus, the problems should be planned in such a way that one week focuses on solving and another on formulating, with the possibility of combining the two proposals, and alternating the types of problems and student records.

The planning should also take into account the different ways in which the children are organized in the room - in a circle, in pairs, in groups, individually - as well as the way in which the children are asked to record the problem, i.e. whether they will solve it orally, or whether they will be asked to draw a picture or text to show how they solved it.

DANTE (1996) highlights the importance of planning in order to achieve children's development in terms of the skills and attitudes involved in Problem Solving. This requires the teacher to reflect on what they are going to do and what has been done at any given moment, which is why they are advised to keep written records in order to progress towards the objectives set and to be able to replan whenever necessary.

4.4.2 Evaluating: the surest way to achieve the expected results

Evaluation according to ABRANTES (1995, p. 27):

> "...is part of the educational process. It is one of the aspects of teaching through which the teacher studies and interprets learning data and their own work in order to monitor and improve the learning process."

The assessment proposed when working with Problem Solving should be continuous, cumulative, diagnostic and formative, with qualitative aspects taking precedence over quantitative ones and the results over the course of the period over any final tests.

"In this way, evaluation points out difficulties and makes it possible to feed, sustain and guide pedagogical intervention." (ABRANTES, 1995, p. 27)

Based on the evaluation proposal, it is necessary to analyze the diagnostic, monitoring and intervention actions and their systematization in relation to certain aspects.

The first of these concerns observation. It is impossible to observe all the students at the same time, so planning and organizing the classroom are fundamental points to guide the teacher's view of the students.

In this sense, organizing a sheet or notebook with the objectives you want to achieve, observing, recording and dating the observations made, so that all the students can be observed at different times, is a simple and practical way to help with the interventions to be made in the planning, or eventually, with some students in particular.

Of course, this takes a bit of work and requires discipline and perseverance on the part of the teacher, but since these are attitudes that we want from our students, it's up to the teachers to set an example. What's more, assessment that depends exclusively on the teacher's memory will be restricted to critical episodes in the classroom, leaving aside a student who speaks little, a different way of approaching a particular idea or activity, or the progress of a student with difficulties.

Another aspect of assessment is monitoring students' development by archiving their work. It's not just a matter of making a folder in which all the student's work is placed, which is usually sent home and over time becomes the preserve of zealous parents or simply disappears.

In this methodological perspective of Problem Solving, we will adopt the processofolio and portfolio, terms used by GARDNER (1994) to identify documentation that not only allows the student to value their production, but also helps the teacher to organize material that gives them and their parents an idea of the evolution of the child's knowledge over the period in which the work was carried out.

In the portfolio process, the records produced by the student over a period of work are archived: texts, drawings and various activities relating to different moments of problem-solving. According to GARDNER (1994, p. 16) :

"The processofolio would be the clearest representative for the child of the durability of their impressions, perceptions and reflections, allowing these elements to be preserved in time and space, which is not always possible through oral language or the teacher's memory. When organizing processfolios, children have frequent opportunities to browse through and look at their work, show it off and interact with their peers. This gives them the chance to be aware of the number of activities they are involved in and the progress they have made."

On the other hand, at the end of a period of work, a portfolio can be put together, i.e. collecting the best productions from among the records made in the portfolio process, those that show something the child has learned new, an original solution selected by the child, possibly with the help of classmates and the teacher. This assessment tool allows students to participate in the organization of their material and to reflect on what it contains, in other words, they are self-evaluating from nursery school onwards. "The portfolio is what the student believes they have done best, what should be valued as their achievements and which can be used by parents and the school community to see the progress of each child." (GARDNER, 1994, p.28)

The processfolios allow the teacher to reflect on which tasks made the most sense, which gave the most effective results, which were confusing and needed further exploration, which students had difficulties or failed to grasp the relationships and concepts involved in the activity. The portfolio shows each child's view of their own work, confirming or not the teacher's observations about them.

In this way, the teacher's records, together with those of his or her students, make up the memory and evaluation of the work carried out, shed light on the way forward, chronicle the life of the class and of each student and legitimize the decisions made.

5 SOLVING PROBLEMS AND RECORDING SOLUTIONS

The primary concern of nursery school teachers is that, when solving problems, children should be able to imagine, construct and seek different solutions in different ways. However, it is believed that it is important for them to realize, at some point, the need to record their solutions in order to communicate ideas, guarantee authorship and think about the path used to solve them.

When developing recording processes with children, it is important to emphasize that "...one of the basic tasks of the school is to train, in all areas of the curriculum, children who are able to read and write independently." (COLL, 1996, p. 37)

Listening, speaking, reading and writing are basic skills for children to learn concepts at any time and serve both to get them to interact with others and to develop a better understanding of the notions involved in a given activity, because according to CÂNDIDO (apud SMOLE, 2000, p.17) "...any means of recording or transmitting information encourages the ability to understand and analyze what is being done." It is in this context of valuing communication in math classes that the various possibilities for recording in problem situations are proposed.

5.1 THE IMPORTANCE OF ORALITY IN PROBLEM SOLVING

At school, orality is the most accessible communication resource that all students can use, whether in Mathematics or any other area of knowledge. Regardless of age or grade, according to CÀNDIDO (apud SMOLE, 2001) "orality is the only resource when writing and graphic representations have not yet been mastered or do not allow the full complexity of what has been thought to be demonstrated."

In this way, the most natural way for most kindergarten students to record what they have done or thought is orally, because almost all of them arrive at school with the ability to express themselves orally. In addition, oral language is a simple, agile, direct communication resource that allows for quick revisions and can be interrupted or restarted as soon as a flaw or inadequacy is noticed.

Opportunities for students to speak in class mean that they are able to connect their language, their knowledge, their personal experiences with the language of the class and, progressively, with the specific expressions and vocabulary of the area in which they are working.

POZO (1998, p. 54) states that:

> "When it comes to Mathematics, whenever you ask a child or a group to say what they did and why they did it, or when you ask them to verbalize the procedures they adopted, justifying them, or comment on what they wrote, represented or schematized, reporting stages of their solution, you allow them to modify previous knowledge and construct new meanings for Mathematical ideas."

In this way, at the same time, students reflect on the concepts and procedures involved in the proposed activity, take ownership of them, revise what they didn't understand, expand on what they did understand and also explain their doubts and difficulties.

Orality used as a resource in Problem Solving can broaden the understanding of the problem and provide access to other types of reasoning. Talking and listening in math classes allows for a greater exchange of experiences between children, expands the mathematical and linguistic vocabulary of the class, and allows ideas and procedures to be shared. Listening to their peers and the teacher, for example, changes their understanding of the statement. Working orally with Problem Solving, even with children who are not readers and writers, is a way of introducing them to this new universe and bringing them closer to mathematical language.

Orality can and should be encouraged when working with Problem Solving: when explaining the solution procedure, when solving in pairs or groups and when solving collectively.

When explaining the procedure used to solve the problem, children can be invited to explain how they thought and clarify any doubts their classmates may have. At this point, they use procedures that don't appear in the written record to explain their reasoning, they create ways of communicating through gestures and expressions that they couldn't include in the drawing or writing alone.

One way of involving the children is to ask them directly if they agree or disagree with a proposal put forward by a classmate. At first, they may simply say yes or no, but over time they start to justify their opinions.

According to CAVALCANTI (apud SMOLE, 2001, p. 127): "Children who don't like to expose themselves during class discussions need a guaranteed space for discussion in groups and pairs. This is a way of ensuring that they can speak and be heard, give their opinion and receive suggestions and points of view from their interlocutor."

Many children will ask questions just to follow what they are hearing; others will give their opinions, judgments and solutions. It's up to the teacher to make sure that everyone understands the task and to try to select problems that are accessible to the class and that are both challenging and don't involve totally new content.

Teachers also need to organize themselves to note down information that will help them plan the next interventions, noting down which students had the most doubts, what kind of doubts they had, which students solved the situation easily, whether or not the class was involved and why. It's not necessary to write everything down in a single lesson; the teacher can plan what they want to observe, choosing one or two items and sticking to them. As the days go by, they will soon have a wealth of information at their disposal, useful in planning and organizing the next lessons.

In small groups or pairs, the children also solve problems proposed by the teacher or created by themselves. Proposing that they solve problems in small groups is a way of ensuring that all the children speak and are heard, and that they receive their opinions. While they are solving the problem, the teacher can go around the groups to observe what each child has done.

According to CAVALCANTI (apud SMOLE, 2001) children's speech indicates what is hidden by the silent consent of the class and even by written records that may not reflect exactly what the child thought when carrying out the task.

To stimulate orality when working with Problem Solving, the teacher can use a variety of proposals, one of which is the *surprise box.* This activity was developed with the students involved in this research and proved the importance of orality.

The surprise box consisted of the students guessing what was inside the box by asking questions about the object. The teacher who hid the object could only answer the class's questions by saying yes or no.

In the first surprise box activity, the children tended to ask questions about what was in the box:

- *Is it a pen?*

- *Is it a sticker?*

- *Is it a pencil?*

By intervening and explaining that they should ask questions about the object, and even giving examples of some questions, they were able to question more meaningfully.

It was observed that the questions were more elaborate during the project and they thought more when expressing their ideas.

- *Is it soft? Yes*

- *Is it white? No*

- *Is it colorful? Yes*

- *Is it a drawing? Yes*

- *Is it about animals? Yes*

- *Is it people? No*

- *I know, is it stickers of people? Yes*

It turned out that they were a little afraid to ask questions at first, because they weren't used to it.

In this first activity, we noticed those students who expressed themselves the most and it was necessary to have a dynamic that would allow the others who were quieter to express themselves as well.

On the other days of the surprise box activity, it became clear that the children were already more open to asking questions, and it was necessary to establish collective rules so that everyone could listen to their classmates and express their ideas.

It was found that the more opportunities children are given to ask questions, the more they will evolve in their reasoning and in their doing and learning of mathematics.

Despite the importance of orality in Problem Solving, it is of great value that students are encouraged to develop other types of recording in problem situations so that they can broaden their communication skills, since different forms of recording have been developed at school to express actions, thoughts and words.

In this sense, varying the recording processes with children is, according to CAVALCANTI (apud SMOLE, 2001, p. 34) "...broadening the possibility of meaning for an idea and allowing the student to acquire increasingly sophisticated modes of expression." For this reason, as well as speaking, we suggest recording through drawing, writing and mathematical language.

5.2   THE PICTORIAL RECORD IN PROBLEM SOLVING

The use of drawings is practically inherent in proposals for work in Early Childhood Education. Drawing, in addition to its artistic aspects, serves as a resource for documenting experiences, sensations and expressing everything that is meaningful to the child.

In addition, according to PILLAR (1996, p. 13), "drawing is a child's first graphic language, their way of expressing on paper their perceptions of the world around them and of what they create, fantasize and desire."

For children who don't yet write, or who already write but haven't yet mastered the mathematical language, drawing is an alternative for them to communicate what they think.

The following example clearly shows that drawing is a way for children to start constructing meaning for the new ideas and concepts they will come into contact with during their schooling.

Although the problem did not require division into equal parts, Felipe showed that he understood division, as he solved the problem using a diagram, thus showing that he understood the meaning of division.

(SCAN FELIPE'S WORK - TURTLE EGGS)

When working with Problem Solving, drawing is important not only for the student to express the solution they have found to the proposed situation, but also as a means for the child to recognize and interpret data from the text. For a student who is not yet a reader, the drawing can serve to support the meanings of the text. In this sense, the drawing in Problem Solving represents both the solving process and the rewriting of the conditions proposed in the statement.

This can be seen in the examples of Rosângela and Elaine, six-year-old kindergarten students.

(SCAN WORK BY ELAINE ALMOÇO AND ROSÂNGLEA CACHORRO)

In these resolutions, the two functions of the drawing that have just been mentioned clearly appear: interpretation of the data and presentation of the solution. Looking at the first record, one can clearly see two distinct parts. The first shows a mother and daughter sitting at a table in the kitchen, which is very detailed, and the second showing the laundry hanging on the clothesline in the laundry room. In the second record, we can see that the child has used drawing to express the solution to the problem.

In addition to these functions, drawing also provides the teacher with clues about the child, how they thought and acted to solve a particular problem, and it provides the child with a means of expressing how they act on the problem, how they express their ideas and communicate.

Drawing for the sake of drawing is not a form of communication, as this involves interaction with other children. In order for this to happen, it is necessary to organize activities that guarantee appreciation of the drawings produced by the children, in other words, making drawing a real vehicle for transmitting ideas. Therefore, "it is important to propose situations in which drawing involves discussion with peers and the exchange of ideas." (ZUNINO, 1995, p. 13)

As the work with Problem Solving develops, the student's repertoire of pictorial resources can be expanded, as long as the teacher has the habit of including other types of representation in their lessons, such as graphs, tables and diagrams.

The children's familiarity with more elaborate representations is revealed in their records, often mixed with drawings and sometimes with texts and even mathematical language.

(SCAN WORK BY CAROLINE DOS GATINHOS)

Six-year-old Caroline drew the problem, showing that she had interpreted it correctly. Look at the cat, the cat and the two bigger kittens. Then she drew the three kittens that died lying in the sky, and the other three that remained alive underneath those that *already* existed. In the end, she carried out the operations according to her thinking and obtained a satisfactory answer.

In addition to the importance of the pictorial representation, the teacher's record of the student's drawing should also be emphasized, as this correct interpretation of the child's drawing was only possible because the child himself explained his solution to his classmates and the teacher wrote it down in his record book.

However, it is not up to the teacher to interpret the child's drawing for the class, but rather to create situations in which the drawing will be used as a language resource, fulfilling an important role as a means of communication. In this way, children will increasingly look for ways to interact with their peers and, over time, will feel the need to include mathematical symbols and signs in order to be clearer, more economical or even faster.

It should also be pointed out that there is no best way to draw and there is no single way either, so you shouldn't teach how to draw to solve a problem or even make drawing compulsory. Instead, children should be encouraged to find the form that is best and most meaningful to them, be it oral, pictorial or written.

5.3   THE WRITTEN AND ARITHMETIC RECORD

On various occasions when kindergarten students solve problems, it can be suggested that they record the solutions they find, or the process of solving them, in the form of a written text.

Writing down an activity helps students to organize their reflections, record their doubts and learning. Thus, this resource in Problem Solving becomes very important when, with students aged five or six, they begin to move towards more systematic records of the solution. According to TEBEROSKY (1996), the text can be written collectively, with the teacher taking on the role of scribe, or individually, if the students are already at the syllabic-alphabetic or alphabetic writing stage.

The text is recorded during or after the discussion of the solutions and, in both cases, serves to formalize the answers validated by the class. In order to organize the text, if it is a collective text, the teacher encourages the students to talk about their solutions, as this conversation will serve as a guiding thread for the writing. The teacher then invites the class to help with the drafting of the text, intervenes by proposing discussions on the organization of ideas and makes sure that the solutions are clear and coherent with the proposed situation.

We can see an example of a written solution that was carried out with the children in this research. The teacher posed the problem and each person said their answer, which was recorded on the board in the form of a list. After everyone had explained their solutions, they were asked to draw and write down their answers.

In this moment of writing, the children used various devices from their literacy process. As the majority of the class didn't know how to read and write conventionally, they tried to copy their answer from the board, noting the initial and final letter of the answer, some words they already knew, and some needed the teacher's help.

The important thing is that even though they copied your answer, they knew exactly what it said, because it was your solution to the problem that was written there.

(SCAN WORK BY ANA AND BRUNO DO LEÂO)

As time goes by, it can be seen that some children are already using different representations - written, pictorial and even numerical - to solve problems, which means that there is a noticeable development of more elaborate forms of representation, an increase in the power to analyze each situation, a greater ability to evaluate results and a great deal of confidence in their own solutions. The following examples illustrate this.

(SCAN ANA AND TAINA'S CHICKEN WORK)

At the beginning of nursery school, from five to six years old, it is not essential to require students to use arithmetic signs to express the solution to their problems. However, getting them to solve problems using different representations, creating their own procedures, should not be seen as devaluing the use of mathematical language or giving it a secondary role.

It is known that, according to POZO (1998), it is not easy for children to express themselves in the conventional language of mathematics, that early demands for operative techniques can inhibit the understanding of a problem and that the acquisition of mathematical language is a slow, progressive achievement, the result of social interactions and many opportunities to express oneself in an original way.

Even so, we shouldn't believe that if students understand the meanings of mathematical concepts and procedures, they won't have any difficulties mastering formal language. This would be tantamount to saying that students who speak well and read easily and comprehensively don't need help to produce written texts, because one day they will write on their own.

To express themselves in mathematical language, children need to master a series of rules and conventions about this language in order to use it properly, which doesn't happen quickly and requires the intervention of the teacher.

According to TEBEROSKY (1996), just as he is concerned that students develop writing skills in their mother tongue, that they appropriate the rules of how that language works and that they have time to do so, teachers must take the same care with mathematical language.

Children resort to drawing, oral or written mother tongue more naturally because this allows them to more easily explain the meanings present in the text, such as words, scenes, information, operations and thus build a mental representation of these meanings.

It should also be noted that, either because of the emphasis placed on communication processes or even because of the relationships that Early Childhood Education students have with other children and adults, it is common for them to feel the need to enrich their representations, which implies modifying their ways of recording problem solutions.

At this point, when the need arises, it is appropriate to inform the child that words can be replaced by a specific sign and to introduce the addition, subtraction and equality signs. The signs can be introduced gradually, in situations where the teacher notices the child's natural involvement, which encourages understanding of the use of arithmetic writing.

When starting arithmetic representations, it is common for students to indicate only the result, only the data of the problem, the data and the result, but without any mathematical signs, the unconventional representation of data and the result, using a mixture of drawings, signs, mother tongue, until they arrive, through the teacher's intervention, at the conventional representation.

At this stage of schooling, students need to realize that there are many ways of solving problems, that they are all valid and that what matters in finding a good solution is knowing what you are doing and why you are doing it.

In the quest to find the solution, the important thing for students to evolve in their representations is to create conditions that allow them to understand the reasons behind the arithmetic representation and to realize that this way of recording the solution to a problem makes it possible to save more time and be more precise in expressing the numerical relationships presented in the problem.

The hope is that, from early childhood, students will realize that learning a language, including mathematics, is not about learning a series of meaningless rules, but about acquiring a degree of communicative competence that allows them to use that language appropriately in the most varied situations.

6   PROBLEM SOLVING IN EARLY CHILDHOOD EDUCATION

Giving students the opportunity to formulate problems is a way of getting them to write down and understand what is important in developing and solving a given situation; what the relationship is between the data presented, the question to be answered and the answer; how to articulate the text, the data and the operation to be used. What's more, by formulating problems, students feel that they have control over the mathematical process and that they can take part in it, developing interest and confidence in problem situations.

When formulating problems, the child tries to think of it as a whole, not just the numbers, a few key words or the question. They become more familiar with and understand the characteristics of problem situations.

As with all text production, the creation of problems must be seen as something challenging and motivating. We need to stimulate children's inventive and questioning capacity, developing a climate of interaction and respect in the classroom, where mathematics can be done through the possibility of questioning, raising hypotheses, communicating ideas, establishing relationships and applying concepts. According to MACHADO (1995, p. 16) "...for students to become effective writers, writing must not be synonymous with tiring, boring or unsuccessful work."

According to MACHADO (1995), the text is not born perfect, it evolves depending on the teacher's interference in proposing different types of texts: collective, group, pair and individual.

Whichever writing resource the teacher chooses, it's important to note that as children gain autonomy, it's possible for them to organize themselves in pairs, small groups or individually in the same activity, according to the teacher's instructions, respecting the way they feel best or can help other children.

In both group and individual texts, the teacher's observations of how their students formulate problems should serve as a tool for immediate intervention with the group or individual student, to enable them to broaden their understanding of the problems and increase their ability to elaborate them.

Working on the production of math problems takes time and patience, because many children are still in the process of testing their writing hypotheses. For this reason, the teacher can sometimes use the text produced to discuss specific literacy issues.

The first proposals for formulating problems must be planned very carefully, since children find it difficult to carry out this task because they are at the beginning of their literacy process.

Students should have contact with different types of problems to solve before proposing that they create their own. It is not a question of solving a large number of problems and, once they have become good problem solvers, starting to formulate them, but rather of giving them a previous experience that allows them to test their hypotheses, get to know and develop models that will serve as a starting point for formulating their own problems.

The first production proposals should be simple, for example: from a picture create a question; from a picture create a problem, from two scenes create a problem; as shown in the example.

(SCANNING ELIAS' PROBLEMS-boy)

It can be seen that the student had to look at the picture and draw some ideas from it that could generate a question. This question can be answered either by looking at the picture or by making assumptions based on what the scene suggests.

The teacher's choice of picture is a task that deserves care so as not to induce too much of what they want the children to ask or answer and also not to be too abstract for children in this age group of five to six years. Ideally, the picture should be of a broad, interesting nature, so as to allow a variety of ideas to emerge.

In these other examples, which follow, it can be seen that at first the children tend to formulate short problems, often without a question, as can be seen in the work of Elias and Bruno, aged five and six. However, as they come into contact with this type of text, their productions become more and more elaborate, as illustrated by the work of six-year-old Tainà.

(SCAN OF ELIAS, BRUNO PAPAGAIO AND TAINA- FLORES PROBLEMS)

It can also be seen that collective work is excellent at the beginning of problem formulation work, as it is from this that children begin to write more.

If the students aren't writers, it doesn't mean that this work can't be done; on the contrary, "proposing collective creation is an excellent strategy for children to develop and solve problems" (MACHADO, p.23, 1995) and, in this case, the teacher can become the scribe of the class and mediate the proposals raised, organizing discussions and guiding the elaboration of the text according to the consensus among the students. The following activity illustrates this very well.

The teacher asked the children to work out a problem for the picture of the train they were seeing. Collectively, the problem flowed. The next day, the teacher brought in a mimeograph of the problem they had worked out to answer the questions they had raised.

When proposing this type of activity to students, the aim is above all to see if they are already getting to grips with the structure of a problem and if they already understand what is essential in its formulation.

It's important to use different types of problems when proposing them to students, so that they can develop different ways of thinking beyond arithmetic.

Because it is so challenging for students, problem formulation should be a space for them to communicate ideas, make points, investigate relationships and gain confidence in their learning abilities. For KRULIK (1997, p. 46) states that "problem formulation is a time to develop notions, procedures and attitudes towards mathematical knowledge."

In addition, the teacher's intervention is essential, as it is the teacher who will ensure that the children gradually get to grips with the characteristics of a mathematical problem, as long as there is room to question the problems produced and reflect on them.

In all the proposed interferences, care must be taken to ensure that students' mistakes are not emphasized as unacceptable flaws that are being shown to intimidate or demotivate them. On the contrary, the aim is for the children to have opportunities to reflect on these mistakes, to point out ways out, to raise opinions, to be able to debate and improve their production.

Another important intervention for students to make progress in producing problems is to create a real intention and an effective target for their productions, which implies a responsibility to make themselves understood, to expose their knowledge and experiences to the gaze of others. It is suggested that a book of problems produced by the students be made, which can be displayed in the school, for parents and the community; it can also be exchanged between classes of the same age group, advertised in the local newspaper, on the school wall, among others. The important thing is that students feel valued for their work and realize that it is an instrument of communication.

In any of the ideas suggested, the children will have to check that the problems are appropriate, that they are of good quality and, if necessary, revise them and work on them, reformulating, reviewing data and improving their writing. This is because when the students' productions are published, the texts must be legible and correct in terms of the proposal made for the formulation and even the spelling of the words. This is because a reader from outside the class group can become, in the eyes of the children, a rigid and intolerant judge, whose opinion can seriously interfere with the students' self-esteem and immobilize them for future work.

Formulating problems is, according to MACHADO (1995), a more complex action than solving problems. In fact, it brings with it solving, insofar as it is necessary to deal with the difficulties of mathematical language, mother tongue and the combination of both according to the purpose of what has been proposed. That's why it's so important to value this other approach to problem solving in early childhood education.

6.1   PROBLEM SOLVING: A PERMANENT ACTIVITY

In order to make the proposal presented so far a reality, it is necessary to make problem-solving a permanent activity.

The permanent activities are repeated in a systematic and predictable way, weekly or fortnightly, and offer the opportunity for intense contact with different types of problems. The distribution of class time demonstrates the importance given to the different contents. By allocating specific, pre-established times to Problem Solving, children are told that this is a highly valued activity. This is one of the benefits of permanent activities.

Ideally, problem-solving should be a constant feature of lessons and every week there should be a problem situation to be explored, discussed and solved. If you want children's critical thinking to develop alongside their language and more mathematical knowledge, they need to be regularly involved in work that enables them to achieve all these aspects of their learning. That's why it's important to do this work frequently in the classroom.

One suggestion that the teacher can use is the *Problem of the Week*. This activity is carried out as follows: on Monday, the teacher poses a very challenging problem and motivates the children to solve it and bring back the solution by Friday, everyone is involved (father, mother, uncle, brother). Some of the teachers at the school in question are doing this and it's working very well, because "the children are starting to think more, and the whole family, the whole class is motivated to solve it", the teachers comment.

The important thing is that these problems often have several answers, or several ways of solving them, and the children are realizing this and, most importantly, they are changing their concept of "problems are just about solving numbers".

Another way to help awaken their thinking and their love of problems is to give them riddles to solve every week (two or three days), both in class and as homework. Some teachers are also applying this suggestion and it is proving to be a success. The students are very motivated to think and all the answers are valued by the teachers.

These are some very simple practical suggestions, but they bring about extraordinary changes in working with Problem Solving.

There are many other practical suggestions for working with Problem Solving which will be explained later in this research, but it is also necessary to let teachers build their own teaching practice, so that they themselves can create other challenging practices; practices which awaken thinking, which lead to the development of creativity, magic and a taste for Problem Solving, both in children and in themselves. This is only possible when Problem Solving becomes a permanent activity in classroom practice.

7 PRESENTATION OF THE PROBLEM-SOLVING PROJECT DEVELOPED WITH THE CHILDREN

In order to analyze the students' attitude and performance when faced with problems, and to prove the importance of practicing Problem Solving, a practical pedagogical project was drawn up consisting of some problem situations, which are suggested when working with Problem Solving in Early Childhood Education. This practical pedagogical project lasted 60 (sixty) hours, in which planning was carried out together with the class teacher, as well as applying the Problem Solving proposal explained throughout this work.

Most of these problems involved heuristic processes. We tried to apply the most varied types of problems to allow an analysis of the strategies used by the students.

This practice focused on metacognitive concepts and aspects (when you think about what you have thought or done) and was carried out in a kindergarten class of five to six year olds at the Francisco Freitas Municipal School, between October and December 2003 (two thousand and three).

The school in question is located on the outskirts of the municipality of Laranjeiras do Sul and operates in its own building in two periods: morning and afternoon. It has four classrooms, a multi-purpose room, toilets, a kitchen and an administration room.

This school has eight teachers, a principal, a pedagogical coordinator and five members of staff to cater for an average of 200 students in Early Childhood Education and Primary Education (1^a to 4^a grades).

The teachers, who are all undergoing ongoing training and with the support of the management and pedagogical coordinators, try to guarantee quality teaching for all students, because the school believes that its fundamental social role is to help form responsible, critical citizens. The teachers are convinced that their primary function is to teach, so they are committed to the students' learning and work on projects to make learning more meaningful.

The clientele served by the school is quite mobile, as according to the school office, the number of transfers received and sent out is relatively high. According to the statistics provided by the school's management, most of the pupils served by the school are from the lower classes, in extremely precarious living conditions, Most of them belong to the São Francisco neighborhood in which the school is located and also to Vila São Vicente, which - according to municipal data and also because I am in close contact with the children who live there - is considered to be one of the most chaotic in the municipality, as it has no basic sanitation among other basic rights of any citizen.

Many students need all the support they can get from the school in order to study, especially when it comes to school materials and even appropriate clothes and shoes, because they don't have enough money to buy them. It is true that a large number of students receive help from the government through programs such as Bolsa Escola, PETI (Program for the Eradication of Child Labor) and others, however, the students in question do not receive this type of help because their age group is below that required by the government.

The class in question, with which I deal almost daily, is very deprived, both in terms of affection and material, and they don't receive much help from their families at home, leaving the school with all the responsibility for their learning. The students are very absent, because their parents have an erroneous view of early childhood education. To alleviate this situation, the school is trying to hold more frequent meetings with families in order to make them aware of the importance of Early Childhood Education and their role at home as educating parents, even if they are illiterate.

With this lack of family involvement in learning, the class teacher, as well as the other teachers at the school, are aware of the importance of their pedagogical action in guaranteeing students the right to learn. For this reason, knowing that they have a great responsibility towards their pupils, they are always looking to reflect on their practice in order to promote meaningful learning and offer quality teaching to all the pupils in this school.

It should be noted that the students who took part in the project already had a basic understanding of quantity and numerals, which had a positive and significant influence on the results obtained.

6.2   SOLVING PROBLEMS: OBSERVING, INTERFERING AND RECORDING

When working with Problem Solving in Early Childhood Education, one of the teacher's roles is to observe the children's actions and interfere so that they can move forward and overcome obstacles, being able to express themselves in different situations, solving, questioning and justifying their problem-solving process.

In addition to observation, records must be kept so that facts and data are not lost during the course of the project.

The problems were always presented to the children with a motivation beforehand, which was done through a song, a story or a game. After the motivation, we moved on to the actual solving process.

The problems were usually read by the teacher or by a child who was already proficient in reading and the other children followed along with the written record they always had in their hands during the reading.

After reading, the children solved the problem orally, and orality is the key point of all the work, in other words, before recording, the children first spoke, discussed, questioned, argued... After all the orality they had experienced, they moved on to pictorial or written records, or both.

Finally, the answers were analyzed collectively and some of the answers were questioned to see if the children were convinced of them and if they really believed in what they had done.

The classroom was usually organized in groups (a maximum of four students per group) or in a circle (for more individual work).

With the work on orality, a constant in class, it was necessary to make agreements with the children beforehand, so that everyone could speak and listen as well. In short, so that everyone could take an active part in the lesson.

During any activity, the teacher's observational gaze allowed for reflection on what was productive, what wasn't, what needed to be repeated, what topic brought the most involvement from the class, which children took part and which didn't. These observations were recorded by the teacher and the researcher to ensure continuity of work and plan interference. These observations were recorded by the teacher and the researcher to ensure the continuity of the work and the planning of interventions.

Some problems were analyzed collectively, and these were also recorded by the teacher or by some children who had mastered conventional writing.

With this method of working, the children learned by asking questions, they learned by arguing, they learned by drawing, they learned by recording, they learned by playing, in short, they learned by solving problems.

At this point, we will report on the work done with the children according to the different types of problems they were given.

6.3   QUALITATIVE analysis OF THE ACTIVITIES CARRIED OUT WITH THE CHILDREN

6.3.1 Charades

When working with charadinhas, o que é o que é and other riddles, it was observed that this is a very interesting resource for non-reading children. This is because these types of texts are very close to their hearts, as they often play riddles at home and their teacher uses them when working with their mother tongue.

The non-numerical riddles allowed all the children to participate, even those who couldn't solve the numerical problems.

The children enjoyed giving their different answers and didn't feel inhibited, since they were seen as games.

At first, the children didn't bother to give a coherent answer to the question, but as time went on, they asked me to read it again so that they could be more coherent when answering.

This development was only possible due to the significant interventions made when checking the hypotheses raised, where the children "thought more clearly" about the answers they gave based on the teacher's questions.

Over the course of the project, the children started creating their own riddles and charades and researching others at home.

The riddles allowed the children to use the same skills involved in Problem Solving: analyzing, seeking to understand, trying to find a solution and checking that the answer was consistent with what was asked. The following solutions clearly illustrate that, in addition, charades could reveal a lot about how children think and their creativity.

(SCAN WORK BY ELAINE GIRAFA, JAQUELINE GIRAFA, ADRIANI GIRAFA, FERNANDA)

6.3.2 Unexpected Situations

These problems were of great importance to these children who were just starting out in the art of problem solving, because they involved more than one possible solution and were rich in possibilities for the students to raise their hypotheses, compare their conjectures with those of their friends, check whether the solution was adequate or not; which required a certain amount of initiative and autonomy on the part of the children.

The children were very involved and motivated to solve this type of problem, which helped them to avoid developing the belief that every problem has only one answer. It also developed their ability to analyze the text and come up with a variety of hypotheses.

Take a look at how some six-year-olds have solved some of the problems.

6.3.3 Problem Situations Involving Division

The first activity was a number problem with more than one solution, which required the children to discover new ways of representing solutions and helped them develop the belief that problems can have more than one solution.

You can see that although the first problem wasn't about dividing into equal parts, some of the children showed that they already had a better understanding and ended up dividing the same amount of eggs per day.

In the other problems involving division into equal parts, most students needed manipulatives to check the veracity of their hypotheses.

It is also worth noting that the numerical problems of division into equal parts sharpened the children's mathematical thinking, and by the end of the project many children were already managing to solve them without the help of manipulable material and were also beginning to realize the need to express themselves graphically.

6.3.4 Problem Situations Involving Multiplication

In these problems, at first the children gave answers "without thinking", in other words, they just said any number.

However, through intervention, through questioning, the children tried to think more and even used their fingers to draw pictures to give a more satisfactory answer.

As a result, they realized the importance of recording and using numerical notation. Through these problems, the children were able to think and express their thinking mathematically, as well as being able to make it clear to others, and this made them value their own thinking and recording, as well as those of their classmates.

What also drew attention in these problems was the argumentation of their answer in relation to that of other classmates who had been different, and the surprise, sometimes, of checking by counting one by one and their answer being "wrong". Their expressions were of amazement and recognition of the error:

- *That's right, it's 12. I got confused.*

Recognizing the error and understanding what caused it was a huge exercise in autonomy and knowledge-building that these problems gave these children.

(SCAN WORK OF PATRIQUE, FELIPE, ELIAS)

7.2.5 Problem Situations Involving Addition and Subtraction

The problems involving multiplication were solved by the children using addition, so they were not addition problems either.

In these problems, the children included quantities using their fingers, but some needed the help of the pictorial record to understand the solution. Other children, in addition to the pictorial record, counted one by one to be more sure of their answer.

Most of the students were able to do the subtraction problem without the help of the recorder and three students were surprised because they solved it orally and when I asked them to explain, they said:

- *I already know, six minus two is four. It's easy.*

- *There were six doves, two went away, four remained, those who don't know have to count on their fingers. Take six and put two down, that leaves four. Is that right?*

- *I already know how to do this kind of math.*

It was observed through oral communication that many of the children were already familiar with operations, what was really missing was solving problems, thinking, questioning and putting forward their hypotheses.

(SCAN WORK BY CAROLINE, JANICE, IN ADDITION TO FERNANDA AND ANA)

Now, the problem shown in the following example, which involved addition and subtraction together, they were only able to solve with the help of the pictorial record and reading the problem slowly, so that they could understand it better and find a satisfactory solution.

They realized that this type of problem involving two operations became a bit beyond what they could produce, as it took a lot of intervention for them to arrive at the correct answer.

(SCAN WORK BY ADRIANI, JAQUELINE,

7.2.6 Logic problems

The logic problems could only be solved because each clue was read in turn and the children were able to eliminate them and find the answer.

It wasn't very easy for them to find the answer, because they wanted to know what it was and not what it wasn't. The intervention and explanation of this type of text was essential. The intervention and explanation of this type of text was essential for the children to be able to solve it.

However, despite being somewhat difficult, the logic problems provided a rich experience for the development of thinking operations such as predicting and checking, hypothesizing, searching for assumptions, analysis and classification.

As the problems were not numerically based, they required the children to use their deductive reasoning much more.

In addition, the logic problems, being motivating, reduced the pressure to get the right answer immediately.

(SCAN WORK BY NERI, VANDERLÉIA, JOSSELIANI)

7.2.7 Problem Solving

Writing problems is a much more complex activity than solving problems, as it requires knowledge of the text to be written, mathematical notions and also having worked with problems before.

When the children were creating their own problem texts, they had to organize what they knew in order to be able to write the text.

The writing was done by the teacher and the students dictated their ideas. Even the quieter ones surprised us with very good ideas for the beginning of the problems.

The text flowed, the children talked too much and came up with very creative ideas, so we had to hold a vote to choose just one idea. This practice wasn't ideal, but there wasn't time to record all the ideas.

The problem text flowed very well and the question?

It took a while for the question to come out, usually they would start telling the problem again or say the answer by example:

- *There were seven left.*

The following examples show that the children's first problems were short texts with few questions, revealing that they had little contact with this type of activity.

(SCAN BRUNO AND ELIAS WORK)

Through a lot of questioning and intervention, they were able to ask not just one but many questions, and as there were so many, they chose just a few to record.

With the elaboration of the problems, the children experienced control over the text and the mathematical ideas, they were no longer just problem solvers but problem proposers and this motivated them immensely in the art of elaborating and solving problems.

7.3 REFLECTIVE ANALYSIS OF THE RESULTS OF THE PROBLEM-SOLVING PROJECT CARRIED OUT WITH NURSERY SCHOOL CHILDREN

Giving students the opportunity to speak in class made them connect their language, their knowledge and their personal experiences with the language of the class and the area of knowledge they were working on.

It was found that when children or groups were asked to say what they had done and why they had done it, and when they were asked to verbalize the procedures they had adopted or to comment on what they had written, represented or diagrammed, this allowed them to modify previous knowledge and thus construct new meanings for mathematical ideas. In this way, at the same time, the students reflected on the concepts and procedures involved in the proposed activity, took ownership of them, revised what they didn't understand, expanded on what they did understand and also explained their doubts and difficulties. This made it possible to propose more and more activities that met the students' needs.

In essence, the dialog in class enabled the students to speak meaningfully, to learn about other experiences, to try out new ideas, to find out what they really knew and what else they needed to learn.

Oral communication also fostered the perception of differences, the students' coexistence with each other and the exercise of listening to each other in a collective learning process, which enabled the children to have more confidence in themselves, to feel more accepted and unafraid to expose themselves publicly. That's why it was so important to provide these moments in math classes where the students could express their ideas and make cognitive progress.

Although some students excelled at speaking, everyone was able to express themselves and even the quieter ones were surprised by their different and creative ideas and this motivated them to express them more often.

As well as speaking, drawing, writing and manipulable materials were other alternatives used by the children to communicate their thoughts.

Drawing enabled the children to start constructing a meaning for the new ideas and concepts they would come into contact with throughout their schooling. This can be seen in the previous examples of the students' activities, where even though they hadn't mastered the technique of division and other operations, they managed to solve the operations required in the problems using the pictorial record, while understanding their meanings.

The students' familiarity with representations was revealed in the records that were mixed with drawings and texts.

These records served as clues as to how each student perceived what they did, how they expressed their personal reflections and what interferences could be made to broaden the mathematical knowledge involved in a given activity.

However, as with any other expression of language, in order to develop, children have to practice pictorial work to master its expression, i.e. the more opportunities they have to draw, the more chances they will have to perfect this type of representation.

Observing the students' records was a clear sign that the children were not only beginning to perceive relationships between different languages in solving problems, but also to take ownership of mathematical writing, giving it meaning.

In addition, during the course of the project, it was observed that some of the children used different representations - written, pictorial and even numerical - when solving problems, which allowed us to see the development of more elaborate forms of representation, an increase in the power of analysis of each situation, a greater ability to evaluate results and confidence in their own solutions. The children's records shown throughout the analysis in this chapter illustrate this.

In this way, this project enabled the children to develop their ability to think mathematically. What's more, this work in which the children devised questions, read and interpreted different types of texts, recorded their hypotheses and solutions using different representations contributed to the children's harmonious development and, in particular, they began to develop the thinking skills that are essential if they are to continue learning and doing mathematics.

8 PROBLEM

According to the teachers' reports, one of the biggest difficulties that kindergarten teachers encounter when working with problem solving with their students is finding varied problems that are suitable for this age group. With this in mind, a small collection of problems was put together, which will be presented below, and given the name Problem Library.

It is suggested, however, that each teacher set up a personal problem library, which can be kept in a box or binder where they will file the problems to be worked on with their students.

The problem library can be set up by the teacher using problems collected from magazines, books and others that he/she prepares or exchanges with colleagues. When organizing your collection of problems, they can be separated according to the examples given below.

This collection is not intended to be complete, but it can certainly serve to guide, support and inspire the work with Problem Solving with students. Remembering that teachers should always look for new problems, because being a teacher is also being a researcher. At the end of the work there are suggestions of books and magazines where teachers can find various problems to work on with their children.

Each of the problem types presented in this chapter are suggestions for the teacher to use in class according to the needs of the students. However, it should be made clear that the different types of problems should not be worked on all at once in the same week. These problems should be solved throughout the school year in a varied and relevant way.

8.1 ADVICE

Charades, what's what, and other riddles are pertinent in Problem Solving work. This is because these types of texts are very close to children's hearts, as their parents often play with them, and they are familiar to children because they are generally used in the literacy process.

What's more, the riddles are not numerical, which allows all the children to take part in solving them, even those who can't solve numerical problems. Proposing these problems in the classroom is a way of getting the vast majority of pupils to take part, because the children like to show their different answers and don't feel inhibited by this, as they are initially seen as games.

Riddles are a type of problem because solving them requires the same skills involved in solving problems: analyzing, seeking to understand, trying to find a solution and checking that the answer is consistent with what was asked.

It has also been noticed that, over time, children start to create their own riddles and research others at home. Here are some examples of riddles and charades that can be worked on with kindergarten students.

J I found a donkey dying of hunger and thirst, looking at food and water, what do you think he preferred?

J Why isn't the giraffe round, white and short?

J Why does the dog eat the bone?

J How do we know where the worm's head is?

J Which insect tells you when it's coming?

J How can you get through the keyhole?

J Why is it that when you throw a coin up, it doesn't fall back down?

J What has five fingers but no hand?

J There were two trucks flying. Then one said: - Stop, trucks don't fly! One fell to the ground, but the other kept flying. Why?

J What color was Napoleon's white horse?

J Why is it necessary to call in three scouts to help the old lady cross the street?

J Five imitation monkeys were standing on a wall. One jumped, how many stayed on the wall?

J There were five birds perched on a tree trunk. A naughty boy hit one of them with a cetra. How many birds were left in the tree?

To find more riddles and charades, teachers can consult the books by BUCHWEITZ (2000) and FURNARI (1994) listed in the references.

8.2   SIMULATING REALITY

Problems of this type are of great importance for beginners in the art of problem solving, because they involve more than one possible solution, do not present numerical data and are rich in possibilities for students to raise hypotheses, compare their conjectures with those of their friends, check whether a solution is appropriate or not, which requires a certain amount of initiative and autonomy on the part of the solver.

Working with these problems helps children not to develop the belief that every problem has only one answer. It also develops the child's ability to critically analyze the text, raise hypotheses, read and interpret different types of texts.

As these problems can be proposed from everyday situations that are very close to the child's heart, the students' involvement and motivation to solve them is great.

It should also be noted that by asking questions about the children's answers, the teacher encourages reflection and helps the children to develop all the skills mentioned above.

Here are some examples of this type of problem:

J Paul doesn't want to cut his hair, it's too long. What would you do if you were Paul?

J The firefighters were called to put out a fire and when they arrived at the scene, they realized that the truck was out of water. What should they do now?

John has a new ball. But he doesn't want to play with it to avoid damaging it. That's why he doesn't even play with his friends and he feels very lonely. What would you do if you were him?

J The teacher asked you to get some material from the school supervisor. But she's not in. What are you going to do to get this material for your teacher?

J You left home in a hurry and when you opened your backpack you realized that you'd forgotten your pencil case at home. What are you going to do now?

J In an assignment you see that your friend is copying you. What do you do?

The teacher can use many other problematizations according to the reality of the class, the neighborhood, the city, etc.

8.3   SITUATIONS BASED ON EVERYDAY LIFE

Everyday situations present many opportunities for developing and formulating problems. According to SMOLE (2000, p. 57) "...content from children's lives or situations they frequently encounter can serve as a context for constructing, inventing and solving problem situations."

For this reason, voting, tidying up the class, controlling book loans, distributing materials, limiting people to groups, planning a job or party, as well as issues involving the people in the class, among others, are good sources of problems.

This way of proposing problems becomes quite natural in various everyday situations at all age levels and requires little prior preparation, but it does require the teacher to be aware of his or her objectives and the involvement of the class in what he or she is proposing.

It's important to note that everyday situations generally involve what the student already knows and present a certain degree of unpredictability, depending on what happens in the classroom. However, these simulations of reality are the closest to what the student is familiar with, so they come across naturally.

Many educators and researchers have dedicated themselves to studying and analyzing games, which demonstrates the undeniable importance of this resource for children's development.

During the game, the child's learning can take place through interaction with the material, the rules and the conflict with the opinions of the other players. In fact, in any game in which there is a conflict of objectives, i.e. the move made by one of the participants interferes with the opponent's decision, each move creates a new problem to be solved, as it alters or confirms the strategy that each of the opponents was using.

Although the game itself is a series of problems for those who play, children don't always understand the strategies, concepts and mathematical procedures that are the objectives of a game. In this sense, the methodological perspective of problem-solving can help to make explicit the important relationships involved in the game and enable children to act more reflexively as they play again and again.

Another way to problematize a game is to interrupt the group or a particular student and ask about some concept or procedure involved, thus creating a problem situation that will be solved orally and which will allow the teacher to observe the students' understanding and their doubts.

An example of this type of solution is the Seven Snakes game, which is organized for groups of two to four children, each of whom must have a sheet containing the numbers 2,3,4,5,6,7,8,9,10,11,12 and two common dice. On their turn, the child rolls the two dice, adds up the points obtained and crosses out the total on their sheet; however, if the total is 7, they must draw a snake on their sheet. If, in one turn, the student gets a total that has already been crossed off, they pass the turn. The person who crosses out all the numbers wins the game and the person who has 7 snakes before crossing out all the numbers leaves the game. This is a very motivating game and allows for several problematizations after it has been played several times so that the students become familiar with the rules and the results obtained.

The following are examples of some questions that can be used to highlight basic addition facts.

S What numbers must come up on the dice for the player to score a 5?

s If all that's left to do is score 10, what numbers must come up on the dice for the player to win?

S What numbers must come up on the dice for the player to draw a snake?

S Why do the numbers that appear in the game start at 2 and only go up to 12?

J A child scored a 6 on one dice. What number came up on the other dice?

8.5 PROBLEMS USING MANIPULATIVE TEACHING MATERIALS

Using teaching materials such as blocks, sticks or tangrams is a routine in many schools.

Although it is known that manipulating didactic materials does not necessarily mean that students understand mathematical meanings and notions, using these resources in activities involving problem-solving is a way of providing students with an opportunity, through reflective action, to understand the mathematical notions and procedures involved in a problem and its solution.

When solving problems with non-reading children, the use of didactic materials helps in the process of simulating solutions and testing hypotheses, allowing students to try, make mistakes, imagine and review their actions.

SMOLE (2000, p. 61) believes that: "such possibilities provide a feeling of security in many students who are unable to master the resources of written representation in problem solving."

Therefore, when solving problems, materials help children to obtain models in order to understand the meaning of the operations or transformations that result from the proposed question.

The use of teaching materials also allows students to use different languages to express problem solving: written, pictorial and symbolic. Discussing these representations and their relationships allows children to understand the meaning of actions and reach a more elaborate level of representation.

There are two most common ways of using didactic material in problem solving: proposing problems to be solved with the material and proposing problems based on the material.

It is emphasized once again that manipulation is not enough for students to develop notions and concepts in mathematics. Nor is this understanding to be found in the material itself.

Any teaching resource should help students to deepen and broaden the meanings of mathematical notions. Therefore, when using this resource with students, the important thing is to see if they attribute meanings to the actions they take and if they reflect on these actions.

Colored cubes can provide good problem situations for students. You could give the students blue, red and yellow cubes and propose the following problem: Always stack three cubes or squares to find the maximum number of towers that are different from each other.

 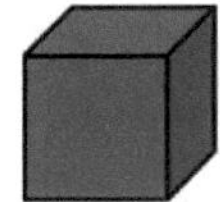 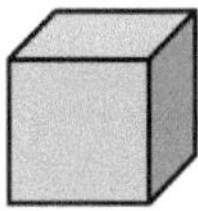

This is a problem whose importance lies in the possibility of students finding more than one answer, organizing counts and records to find the solutions and analyzing each tower to see if it isn't a repeat of a previous one. What's more, it's a problem that allows the answers to be solved and analyzed collectively.

The tangram is also a very interesting material for working with this type of problem, as you can tell the legend of the tangram, which motivates students a lot. Here are some examples of proposals that can be worked on with students using the tangram.

S Use 3 tangram pieces to form a boat.

S Use 2 tangram pieces to form a house.

S Use 4 tangram pieces to form a house.

S Form a square using 4 tangram pieces.

S Form a rectangle using 3 tangram pieces.

S Build a trapezoid with 4 tangram pieces.

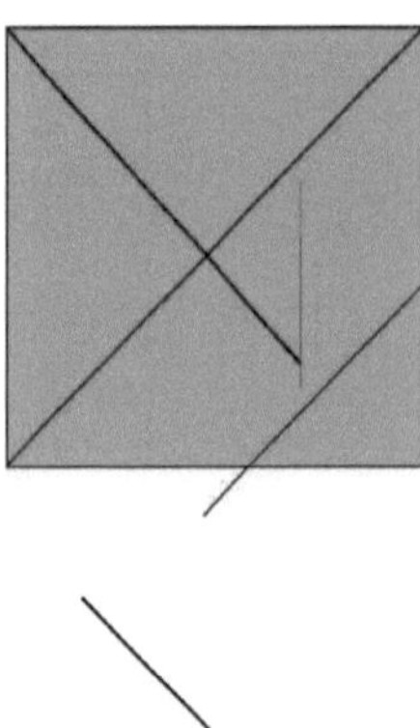

Sticks, a very common material, are excellent for proposing very challenging problem situations to students, take a look:

J Move 2 toothpicks so that there are only two squares.

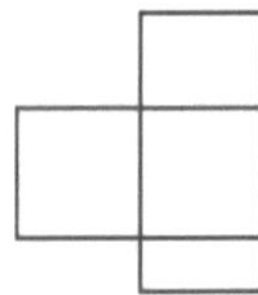

J How do you make 2 triangles with 5 sticks?

S How many different outfits can we make with the items of clothing shown below?

In addition to these examples, there are many other problematizations that the teacher can make, emphasizing that discussing the answers is essential to developing the students' heuristic thinking.

8.6 SOLVING TEXT PROBLEMS

Once students have developed the skills to understand, solve and represent problems proposed orally, whether they are numerical or not, they can be asked to solve problems presented in writing.

Proposing the written problem and questioning the class orally, as is often done when discussing a text, helps with the initial work with written problems. However, care should be taken not to solve the problem for the students during the oral discussion.

If the teacher provides each student with a sheet of paper with the problem written on it, the teacher can also: ask the students to find and circle certain words; choose a word from the problem and ask the students to find others in the text that begin or end with the same letter or the same sound and you can write them down in a list; write the text of the problem on the board without any words and ask the students to look at their texts and find out which ones are missing and then write those words on the board; among many other ideas that will certainly come up when working with the problem text.

When working on text problems with students, it's worth taking care that they are not always of the same type or exclusively numerical. According to SMOLE (2000, p. 92), working with a variety of text problems should be done:

> "...from Early Childhood onwards, encourage the development of different ways of thinking beyond arithmetic, stimulating diverse reasoning, developing varied solution strategies and enabling the need to understand a situation, consider the data provided, collect additional data, discarding irrelevant data, analyzing and drawing conclusions from the data, devising a plan for solving and, finally, solving and checking the coherence of the solution, which are common procedures both in understanding different types of texts and in math problems."

However, it is also possible to propose problems involving other types of texts, as the following examples show.

8.6.1 Rhyme problems

Rhyming problems are characterized by their poetic language, similar to the poems and nursery rhymes that children often hear and recite from an early age. Their musicality allows them to break the conventional structure of math problems and stimulate their interpretation in a variety of situations.

Here are some examples:

S I baked some cakes

6 coconut and 1 English

When they're ready

How many do I eat at once?

S Paula has a rose

Three violets, 2 jasmines

Your ______________ flowers

None for me.

S There are 12 stars in the sky

All of them in rows

One is mine, four are yours

The others are Mariazinha's.

S Uni, duni, tê

Two colored ice creams

Three brigadeiros

My   sweets to you.

S Oba! It's a party in the countryside!

Eight children are playing.

The accordion player

Everyone runs without getting tired.

But...

When the music stops, the rush is general.

They all want to sit down

How many will be left, only 4 chairs to rest on?

There are seven boys.

On the pitch, very excited, they play three against three.

From the outside, very upset,

Ronaldinho waits his turn.

Look, another one has arrived!

Now they are!

Everyone will play!

in Zé 's team,

in Mané   's team.

8.6.2 Logic problems

These problems are not based on numbers, but require deductive reasoning and provide a rich experience for the development of thinking operations such as predicting and checking, hypothesizing, searching for assumptions, analyzing and classifying.

The trial and error method, the use of tables, diagrams and lists are important strategies for solving logic problems. In addition to the requirement to use one of these strategies to solve them, they encourage more analysis of the data, encourage reading and interpreting the text and, because they are motivating, reduce the pressure to get the right answer immediately.

(SCAN PICTURES AND THEN TYPE IN THE PROBLEMS)

8.6.3 Problems Involving the Four Operations

When proposing problems of this type, there should be no preoccupation with only addition or counting problems.

Similarly, it is interesting that numerical problems have more than one solution, which ultimately requires children to discover new ways of representing solutions and prevents them from developing the belief that numerical problems have only one solution.

It is also worth noting that number problems can serve as a way of encouraging students to feel the need to express themselves graphically and to improve their representations. Below are some examples of problems that can be worked on with kindergarten students.

They are three little houses made of clay. In each house there is a bird hatching 2 eggs. How many birds are there?

S Two men were walking along a road. Each man was carrying 2 donkeys, each donkey was carrying 2 baskets. In each basket there were 2 chickens. How many chickens were there?

S How many sandwiches can I make with 12 slices of bread?

J In a forest there was a tree with 2 branches. Each branch had 3 nests and in each nest there were 2 little eggs. How many eggs are there in this tree?

J He-man and Skeleton are great enemies. He-man wants to destroy Skeleton's staff, but the staff shoots laser beams three times a second. How many shots does He-man receive in 4 seconds?

J In an aquarium there were 2 little fish. In another tank there were 3 little fish. Suddenly the tank with the 2 fish broke. The other fish said: - Come on, jump in, if you don't want to die the solution is to stay together! How many fish stayed together?

J Mico and Tico are two monkeys who are crazy about bananas. Wandering around their home at the zoo, they found 8 bananas that the zookeeper Pedro had left there. After much discussion, they decided to divide up the bananas so that the two of them got the same amount. How many bananas will each of them get?

J Rodrigo has 2 rabbits: Juquinha and Piteco. Rodrigo bought 10 carrots and wants to give them to his bunnies so that they both get the same amount of carrots. How many carrots will each rabbit eat?

J Fàbio had 7 carts and got 3 more from his uncle.

- How many cars does he have now?

- Fàbio's carts have 4 wheels each. If Fabio decided to count all the wheels on his carts, how many wheels would he count?

- On Monday, Fàbio decided to play with his cars in the sandpit and lost 4 cars. How many cars did Fàbio return home with?

S There were 5 passengers in a bus. At one point, 1 passenger got off and 2 got on. How many passengers will continue?

S There are 5 children playing. They all have to shake hands. How many handshakes will be given?

S On Mother's Day, Rui's mother got 5 flowers from his father and 3 flowers from his children. Draw all the flowers Rui's mother got that day on the vase.

1.1.4 Unsolved Problems

Working with this type of problem breaks with the concept that the data presented must be used to solve the problem and that every problem has a solution. It also helps to develop the student's ability to learn to doubt, which is part of critical thinking.

Examples:

J A boy has 3 carts with 4 wheels each. How old is the boy?

J How can I divide 2 cats equally between 3 people?

J Draw a rectangle using only three lines.

J There are five books on a shelf. How many of them are math books?

1.1.5 Problems with more than one solution

The use of this type of problem breaks with the belief that every problem has only one answer, as well as that there is always one right way to solve it, even when there are several solutions, one of them is correct.

Working with problems with two or more solutions makes students realize that solving them is a process of investigation in which they participate as thinking beings and producers of their own knowledge.

Take a look at some examples that can be used in the classroom with preschool children.

J Two baskets have 9 oranges in them. How many oranges are in each basket?

J André and Tuca have 11 marbles together. How many marbles does each of them have?

J Mrs. Eulalia is going to distribute 7 apples to her 3 children, all of whom will receive the same amount. How many apples will be left with Mrs. Eulalia?

J You and I have 6 reais together. How much money do we each have?

8.7   DEVELOPING PROBLEMS

At first, you could make proposals to help students understand the form of a problem text and what is essential in its formulation.

You could also ask students to invent problems in the following ways.

8.7.1 Creating Another Question

The idea here is that students, after solving a problem proposed by the teacher, can recognize the data, the situation created, the questions already asked and create a new question for it.

One example is the problem that was worked on with the children in the survey. Look at the questions they came up with after solving the following problem: Claudia had a cat and a kitten who was pregnant. One day, eight kittens were born. How many kittens did Claudia have?

What were the names of the cat?

What will Claudia do with the kittens?

How many eyes do all cats have?

And if one dies, how many will remain?

And if Claudia gives 2 kittens to our class, how many kittens will she get?

8.7.2 Creating a Similar Problem

In this activity, after the students have solved a few problems of the same type, the teacher proposes that they create a similar problem collectively or in small groups.

Once the proposal has been made, the teacher organizes the class so that everyone can present their ideas, encouraging them to speak, questioning when necessary and recording the text according to what the class has proposed, taking care to discuss the writing of the words, the order of the ideas, where the question goes, etc.

In fact, the process of working out a problem collectively is similar to that used to record the children's solutions in writing or produce any other collective text. The difference in the process comes when it is necessary to discuss with the students aspects that are characteristic of problems, such as always having one or more questions to be answered.

As you see your students become familiar with and skilled at formulating problems, you can vary the proposals.

8.7.3 What's the question?

In this suggestion, the teacher gives the students some data, or an incomplete problem, and asks them to ask a question involving the data.

Students are encouraged to create questions for the problem.

It is interesting that, after listening to and discussing the questions with the class, the teacher writes the text down on the board along with the respective questions and then reads the problem out loud to the class.

8.7.4 What's the problem?

In this case, students are encouraged to formulate short texts for a problem based on data provided by the teacher, or even based on an answer. Examples:

S Four carts and a truck.

S The answer is 8 reais. What's the problem?

S Dog, four legs, three dogs. What's the problem?

There are 6 little fish and 3 aquariums. What's the problem?

S One cage, 5 birds. What's the problem?

Another possibility is to give the students a sheet of paper with drawings or cut-outs and, given an answer, ask them to work out a problem that has that answer.

8.7.5 We have a problem

In this proposal, the teacher divides the class into groups of four and asks them to carefully observe the class or other things indicated by the teacher and look for data to ask a problem or a question.

The group that comes up with a problem says: *We have a problem*. At this point, everyone stops to listen and solve the problem, which will also be recorded in writing.

8.7.6 Creating a Story

The aim of this activity is for students to be able to work out problems with longer, more complex texts, which is why the designation says that they have to create a story, which is a familiar text genre. They know that it has a beginning, middle and end. However, as the aim is for them to understand what a problem is and how it is developed, the students should be told that the story should end with questions which they will answer later. The story can be created from a picture, a theme or a subject, as long as it interests the students at the time.

As the formulation proposals progress, and if the teacher leads the students to take an interest in the theme or the picture that is generating the problem, it is normal for the texts to become more complex.

It is advantageous for children's texts to have open questions, or too much data, and to be longer than "usual", as this prevents them from developing inappropriate beliefs about problems and their solutions, such as that all the data in a problem is in the text and will be used to solve it, or that problems always have a solution.

8.7.7 Problems from a Picture or a Print

For non-reading children, one of the simplest ways of presenting and formulating problems, whether numerical or not, based on a textual representation are pictures. According to SMOLE (2000, p. 55) "...the questions are formulated in such a way that the pupils need to read the information contained in the picture and select the information needed to answer the questions."

The students, in small groups, will have to find questions they can answer by looking at the picture. First, the questions are produced orally, then they are written down with the teacher's help.

Questions that require a more in-depth treatment, of a deductive nature, will be proposed by the teacher himself to the class, who will ask the students to answer them.

The teacher should not focus on correcting the answers, but should encourage the students to explain what they have observed and what reasoning they have used to arrive at their answers.

The discussion of the questions is important because, among other things, it shows that they are of a different nature because they require different ways of thinking: simple reading of the picture or text or more deductive activity.

Likewise, they need to realize that certain questions cannot be answered if we limit ourselves to the information provided by the image. The teacher can highlight the questions which, although related to the event presented by the image, cannot be answered from the picture.

Although it is a rich and simple resource for proposing problems, some care must be taken when selecting the images to be used. Naturally, the first is to select pictures that don't contain scenes depicting embarrassing, violent or prejudiced situations. The second is that the scene depicts something motivating, either because it is close to the children's experience, or because it is an amusing situation or related to a specific interest of the class.

(SCAN PICTURES)

8.8   WHAT TO DO WITH THE PROBLEMS STUDENTS FORMULATE AND SOLVE

One of the aspects emphasized in this problem-solving proposal is the importance of communication as a way of enabling students to broaden their understanding of mathematical concepts and procedures. When proposing that students create problems, this principle must not be forgotten.

The problems formulated by the students must be solved by the class (or by other classmates).

The problems can also be put on a poster and displayed in a visible place, reproduced for all the students, or even form a small book of problems prepared by the class throughout the year.

These procedures, which create a reader for the problem, encourage the students to produce better and better problems and show the children that, just as with their mother tongue, the texts produced in mathematics have the function of exposing, recording, marking a position and, for this reason, it is important that they are clear, precise and well prepared. (SCAN COVER OF TWO STUDENTS' WORK)

9   FINAL CONSIDERATIONS

Looking at Problem Solving in a new way, it comes to be seen as an objective of Mathematics teaching that provides a differentiated positioning in the face of challenging situations. Solving each problem is a moment when children have the chance to try to find their own way, develop arithmetic relationships in a contextualized way and reflect on mathematical operations.

On the other hand, it is necessary to ensure that they have access to mathematical language through successive approximations, which should start from nursery school through permanent work with Problem Solving. As you can see, this is not simple, it takes time and depends a lot on good planning. Working with the methodological perspective of Problem Solving does not include occasional experiments, nor does it allow for improvisation or a lack of clarity about mathematical knowledge and the appropriate way to use the methodology.

When working with Problem Solving in Early Childhood Education, the main objective is to enable children to develop their ability to think mathematically and analyze their reality by making different types of relationships. To do this, they need to be encouraged to actively engage in new situations. In this sense, it is believed that working with different explorations and reformulations, seeking to develop interest in the problem, exploring its language, encouraging and challenging children, contributes significantly to making them much more autonomous and capable of tackling the problems proposed without fear or trepidation.

Therefore, the Problem Solving proposal presented in this paper indicates that more important than giving the correct answer to the proposed problem is for the child to be able to explain how they arrived at the answer, justify their strategies and analyze whether the chosen path is the only one that leads to the solution. By getting used to doing this kind of analysis, they will be able to develop more flexible forms of reasoning and apply the strategies they have built up to new situations.

In this proposal, the educator is not just a problem solver, but a problem solver, the one who gets the students to think, to question, to look for different solution strategies, in short, who through their pedagogical action enables children to have new opportunities to learn Mathematics in a challenging and fun way.

It is therefore hoped that this work will help early childhood teachers to organize their work in mathematics classes in such a way as to encourage students to learn not only mathematical concepts and ways of thinking through problems, but also to develop reading, writing, interpreting and producing texts.

By applying the Problem Solving proposal in the classroom, the aim was to break down inappropriate beliefs about what a problem is, what it means to solve problems and, consequently, what it means to think and learn about mathematics. Choosing to work on a variety of problems enabled the children to change their attitude towards problem solving. The aim was for them to develop a critical sense, a spirit of investigation, creativity and autonomy, becoming capable of facing, observing, discussing and deducing challenges, persevering in the search for possible solutions.

It is hoped that the problem-solving work carried out with the children has helped them to continue learning, confident in their ways of thinking and confident to dare and create.

This research analyzes and proves the importance of working with Problem Solving as a methodological perspective in the teaching and learning of Mathematics in Early Childhood Education, with children aged five to six. And to show that they should be in permanent contact with mathematical ideas and that this is possible through working with problem-solving that is challenging and motivates learning.

yes
I want morebooks!

Buy your books fast and straightforward online - at one of world's fastest growing online book stores! Environmentally sound due to Print-on-Demand technologies.

Buy your books online at
www.morebooks.shop

Kaufen Sie Ihre Bücher schnell und unkompliziert online – auf einer der am schnellsten wachsenden Buchhandelsplattformen weltweit! Dank Print-On-Demand umwelt- und ressourcenschonend produziert.

Bücher schneller online kaufen
www.morebooks.shop

Printed by Books on Demand GmbH, Norderstedt / Germany